What's [illegible] on Earth?

MW00570296

by LISA JO RUDY
With the Editors of TIME For Kids

Table of Contents

Introduction

How many kinds of animals live on Earth? Scientists think there may be about 13.5 million. In fact, scientists are finding new forms of life each year.

Why did it take so long to find these animals? Some were at the bottom of the ocean. Some were inside hidden caves. Others were deep inside forests on tall mountains. Is there more life on Earth to be discovered? Yes, there is!

This map shows where scientists have found new animals.

What Is a Species?

Scientists sort all living things into groups. Plants make up one large group. Animals make up another. The animals group is divided into two big groups. These are animals with backbones and animals without backbones. Humans, birds, and fish are some animals with backbones. Jellyfish, insects, and worms are some animals without backbones.

The groups into which animals are divided get smaller and smaller. The smallest group is called a **species.** In this group animals are very much alike. They can also produce more of their own kind. Human beings and American robins are examples of different species.

American robins

Look at all the groups the American robin belongs to!

Chapter 1

Something New at the Bottom of the Ocean

It is dark and cold at the bottom of the ocean. This is where scientists have found **deep-sea vents.** These are cracks in the ocean floor. They shoot out super-hot water and chemicals. It is surprising, but many animals call these vents home!

Most of these animals were discovered in the past 25 to 30 years. Before that, no one knew about the vents. Now scientists have tiny submarines. They use them to explore the deepest parts of the ocean. Each year, scientists dive deep to find amazing new creatures!

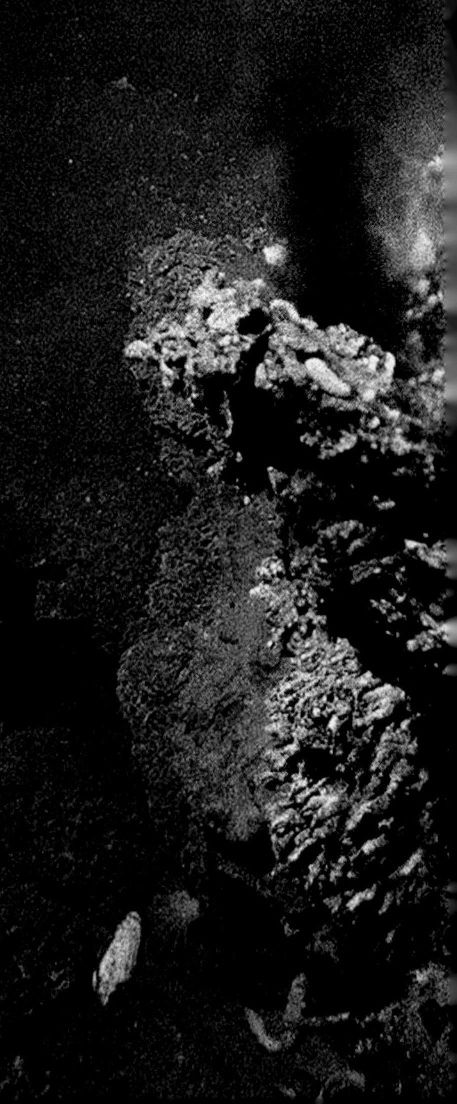

Hairy Crabs from a Strange, Dark World

Dr. Joe Jones is a scientist. He studies the deep sea. In 2005, Dr. Jones sailed with other scientists to an area near the South Pole. Three of the scientists took a tiny sub to the ocean floor. There, they found a deep-sea vent.

Dr. Joe Jones

Near the vent they saw white crabs with hairy arms and legs. Using the submarine's robot arms, the scientists collected one of these crabs. Dr. Jones e-mailed a picture of the crab to an expert on vent crabs. "It's nothing I've ever seen before!" the expert replied. The scientists named it the Yeti crab for the hairy Yeti creature of legend.

Alvin, the mini-sub the scientists used to explore deep-sea vents

Yeti crab

Is It a New Species?

Dr. Jones thought the crab was a new animal species. To make sure, he ran a test. He checked the crab's **DNA**. This is a chemical that affects how living things grow. Every species has its own DNA.

Dr. Jones compared the Yeti crab's DNA to the DNA of other crabs. The DNA was different. The Yeti crab was a whole new kind of animal!

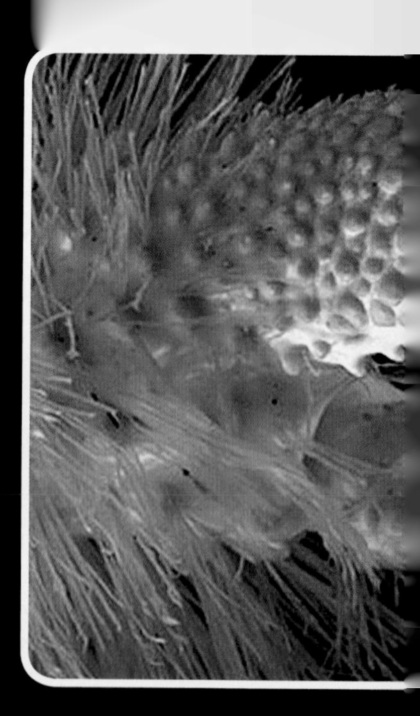

Joe Jones, Undersea Explorer

Here's what Dr. Jones says about diving to the bottom of the sea in a tiny submarine.

Question: What is it like to explore in a mini-sub?

"Only three people can fit inside. There's no room for seats, so everyone sits on cushions to protect them from the freezing cold metal at the bottom of the sub. Each person has a six-inch (15-centimeter) window."

Question: How do you bring back animals from the bottom of the sea?

"On the outside of the sub are two robotic arms. The pilot 'flies' the sub and works the arms to pick animals up. Then the robot arms place the animals in special boxes with lids. We examine the animals when we get back to the surface."

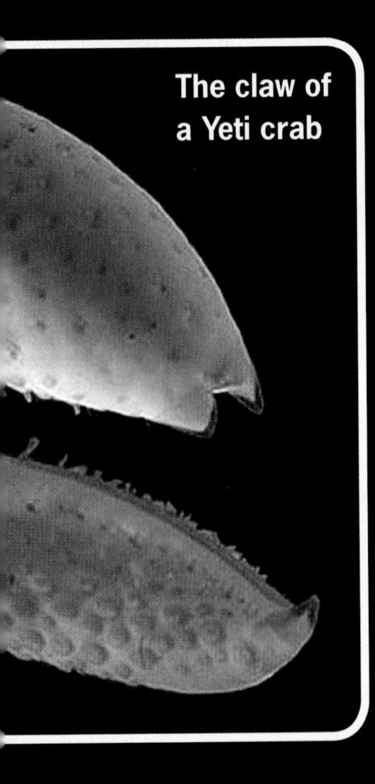
The claw of a Yeti crab

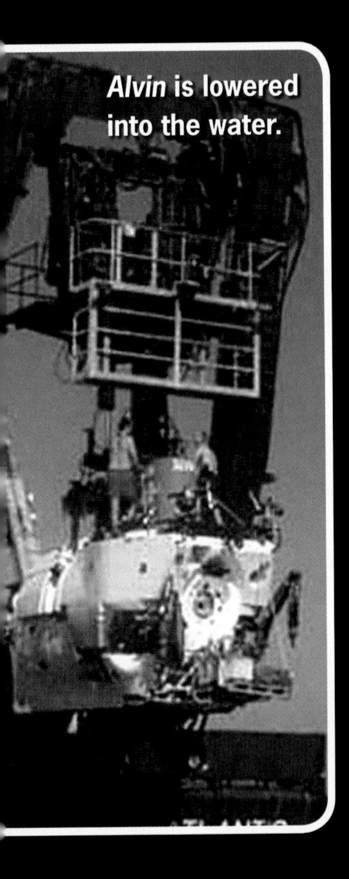
Alvin is lowered into the water.

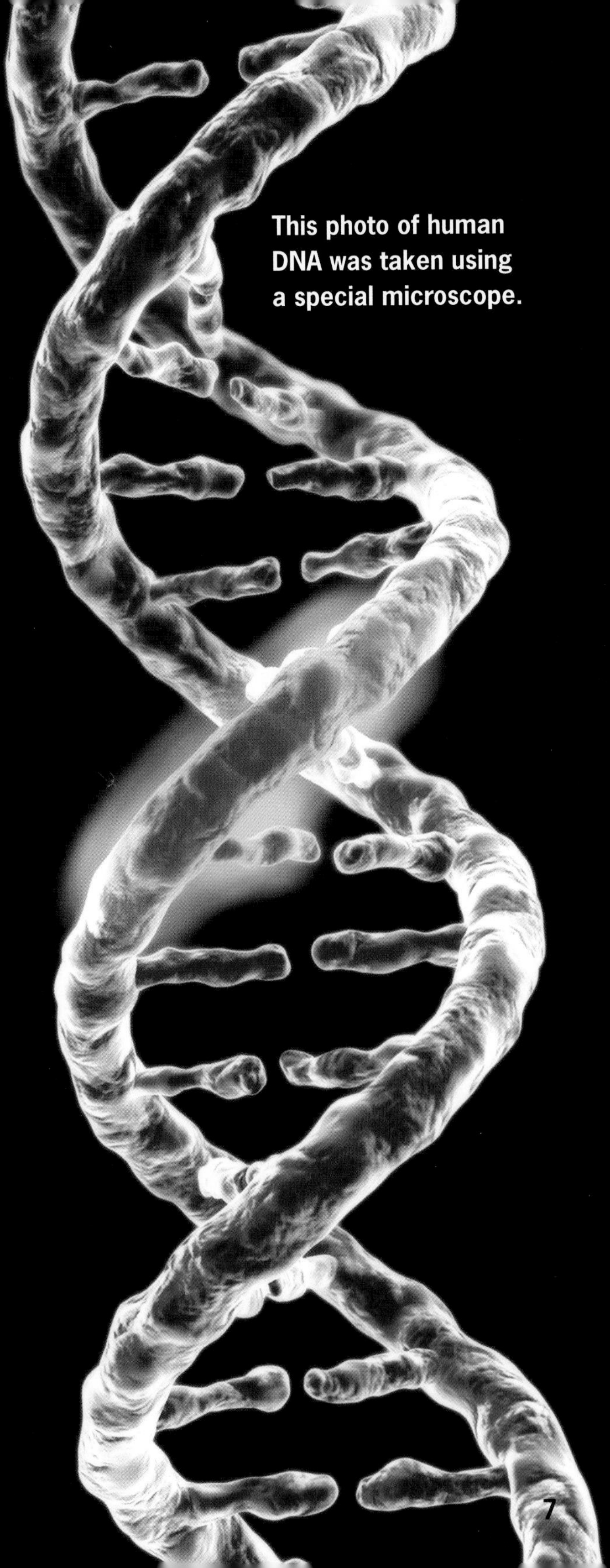
This photo of human DNA was taken using a special microscope.

Chapter 2

Something New in the Mountains

In 2005, Bruce Beehler, a bird scientist, came to the island of New Guinea (GIN-ee). He and a team of scientists hiked into the Foja Mountains. These mountains are covered with **rain forests**. The scientists were looking for new species. In just two weeks, they discovered dozens of animals they never knew existed.

The first new species was a bird. It had an orange face and ate honey. It was the first new bird species discovered in New Guinea in more than 60 years! The scientists called it the smokey honeyeater.

Smokey honeyeater

The team also discovered more than 20 new frog species. One tiny frog was only 14 millimeters (about half an inch) long.

They also found four new insects on the trip and many new plants. Everywhere Dr. Beehler looked, he saw amazing things he had never seen before.

One of the 20 new frog species found in New Guinea

Creepy Cave Critters

Movile Cave in Romania was discovered in 1986. Before that, no one had ever been inside it. So far, scientists have discovered more than 33 new species of plants and animals inside. They have found flat worms, round worms, scorpions, leeches, pill bugs, centipedes, millipedes, and spiders. Before 1986, most of the species had never been seen by humans.

Why are so many new animals living in one cave? Nearly six million years ago, the climate of Romania changed. Instead of staying warm all year, it began to get cold in the winter. Animals died. But animals that lived in the warm cave survived.

Movile Cave has no light and very little air. How can animals live there? Over time, the animals developed new **adaptations**. These traits have helped them survive in the harsh **environment**.

The animals in Movile Cave are a lot like the animals that live near deep-sea vents.

- They are lightly colored.
- They are blind.
- They use other senses to find their way in the dark.
- They can live by eating tiny animals and fungus.

A "new" centipede (above) and spider (right) found in Movile Cave, in Romania

Chapter 4

More Mammals in the Mountains

The Annamites are tall mountains in Asia. They are covered with thick, wet forests. No people live nearby. But inside these mountains is a hidden world of living things.

In 1992, scientists from Vietnam found a new species of large **mammal**. It was living in the mountains. It was called a saola. The saola looked like an antelope. It was the first new mammal found in the Annamites.

The Annamite Mountains divide the countries of Cambodia, Laos, and Vietnam.

The saola is a deer-like mammal discovered in 1992.

Three years later, scientists in Laos discovered the body of a mammal. They were told it came from deep inside the mountain forests. The scientists went into the forests. They set up a special camera. When it sensed movement, the camera took a photograph. It took pictures of the new mammal. Scientists called the mammal an Annamite striped rabbit.

In 1998, scientists returned to the mountains. Using the same kind of camera, they took pictures of a Javan rhino. People living in the area had seen the rhino. Yet no one had ever taken a picture of it. The scientists also discovered a new species of bird called a golden-winged laughingthrush.

The golden-winged laughingthrush was discovered in 1998.

A special camera snapped this picture of a Javan rhino.

Conclusion

Much of Earth has still not been explored. Most of Earth is under the ocean. Big chunks of Earth are frozen at the North and South Poles. Scientists have just begun to explore new jungles, forests, mountaintops, and deserts. There will be plenty of new species for them to find.

Animal Species by the Numbers

- Scientists think that about 13.5 million species of animals live on Earth.
- We have found only about 1.75 million species so far.
- Each year, we discover 15,000 to 20,000 new animal species.

It could take another 1,500 to 15,000 years to find all the species living on Earth!

A pillbug found in Movile Cave

This bird, not seen for more than 100 years, was rediscovered in the Foja Mountains.

Glossary

adaptation (ad-uhp-TAY-shuhn) a trait that helps a living thing survive in its environment *(page 11)*

deep-sea vent (DEEP SEE VENT) a crack in the ocean floor that shoots out super-hot water and chemicals *(page 4)*

DNA (DEE-EN-AY) the substance that gives living things their features *(page 6)*

environment (en-VYE-ruhn-muhnt) the things that make up an area, such as land, water, and air *(page 11)*

mammal (MAM-uhl) an animal that has fur or hair, feeds its young with milk from the mother's body, and is warm-blooded *(page 12)*

rain forest (RAYN FAWR-ist) a dense forest that gets a large amount of rain during the year *(page 8)*

species (SPEE-sheez) a group of similar organisms that can produce more of its own kind *(page 3)*

Index